ZOOM
Space
BLACKBIRCH® PRESS
THOMSON GALE
San Diego • Detroit • New York • San Francisco • Cleveland
New Haven, Conn. • Waterville, Maine • London • Munich

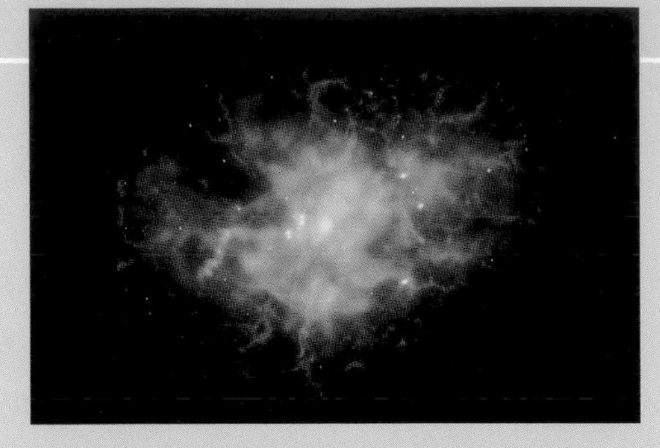

# CONTENTS

# LET'S ZOOM!

When you use the zoom feature on a camera, you can bring an image from the distance into close-up focus without moving yourself. For example, you can capture the image of a butterfly on a leaf while keeping your distance. This book works in exactly the same way.

Imagine you were able to point a camera at the whole Universe. The illustration on pages 6 and 7 shows what you would see in your viewfinder. Now zoom in to one part of the scene. You'll see much more clearly what those spots of light look like in close-up (they're galaxies). Keep zooming, and a few of that galaxy's billions of stars will come into view. And so on, until you've focused on one particular star with its family of planets, one very special planet (our Earth), the surface features of Earth, the rocks of which it is made . . . and eventually the minute building blocks, called atoms, which make up the rocks—and everything else in the Universe.

It's a fascinating journey through space, yet you will not have to move one millimetre! And you will discover some amazing things about space that only this incredible *zooming* book can show you . . .

# UNIVERSE

The universe is everything we know. All matter, from the tiniest grain of sand to the most gigantic star, belongs to the universe. It even includes empty space.

Scientists think the universe began in an incredible explosion that happened about 15 billion years ago. During this event, called the big bang, all matter, energy, and space were created. The universe has expanded enormously since the big bang.

Matter in the universe is not evenly spread out; it is clustered in a net-like pattern. The "holes" between matter are vast empty spaces—voids. The clusters of light are not stars, however, but galaxies.

ZOOM DOWN IN TO A SUPERCLUSTER OF GALAXIES

THIS CLUSTER IS ONE OF BILLIONS IN THE UNIVERSE

# GALAXY CLUSTER

The universe is made of billions of galaxies, collections of stars. But galaxies are not found on their own, spread out across space. They are clumped together in clusters, each crowded into superclusters. These giant "clouds" of galaxies span hundreds of millions of light years. Superclusters tend to be stretched into long "strings" that link together across the universe. In between the superclusters lie the vast empty spaces, the voids. Our own galaxy, the Milky Way, belongs to a cluster called the Local Group. It consists of about 30 galaxies.

The Local Group contains different types of galaxies: spirals (like the Milky Way), barred spirals (a little different from the spirals' catherine-wheel shape), ellipticals (shaped like ovals), and irregulars (no obvious shape). The nearest galaxy to Earth, the Large Magellanic Cloud, is 170,000 light years away.

**ZOOM IN TO THE SPIRAL-SHAPED MILKY WAY GALAXY**

***Irregular***

***Spiral***

THE MILKY WAY GALAXY BELONGS TO A CLUSTER

# GALAXY

The Milky Way galaxy is a huge, flat spiral of stars. It is named after the misty band of stars in the night sky (we only see a side view of one of its spiral arms). The Milky Way contains 200 billion stars and measures 100,000 light years across. It spins at 825,000 miles (250 km) per second.

Nucleus

Spiral arm

The Milky Way galaxy has a bulge at its center, called the nucleus. This is where older, red stars are concentrated. Four giant arms spiral out from the nucleus. Younger blue stars are found in these arms, along with clouds of gas and dust, where new stars are forming *(see page 24)*. About half-way out from the nucleus are the middle-aged stars, mostly yellow and orange.

ZOOM DOWN IN TO ONE OF THE SPIRAL ARMS

THESE STARS ARE SOME OF BILLIONS IN THE GALAXY

# SPIRAL ARM

Stars produce energy (including heat and light) that they radiate in all directions as they shine. These hot, spinning globes of gas vary enormously in size. They also vary according to the amount of energy they give off. Many stars may have planets orbiting around them—some perhaps home to living things.

ZOOM IN TO OUR SOLAR SYSTEM

The shimmering, multicolored region in this illustration is known as a nebula *(see page 24)*. It is a vast cloud of dust and gas—the remains of dying stars. Some of the galaxy's nebulae take on spectacular shapes. Nebulae provide the "raw material" from which new stars will begin to form.

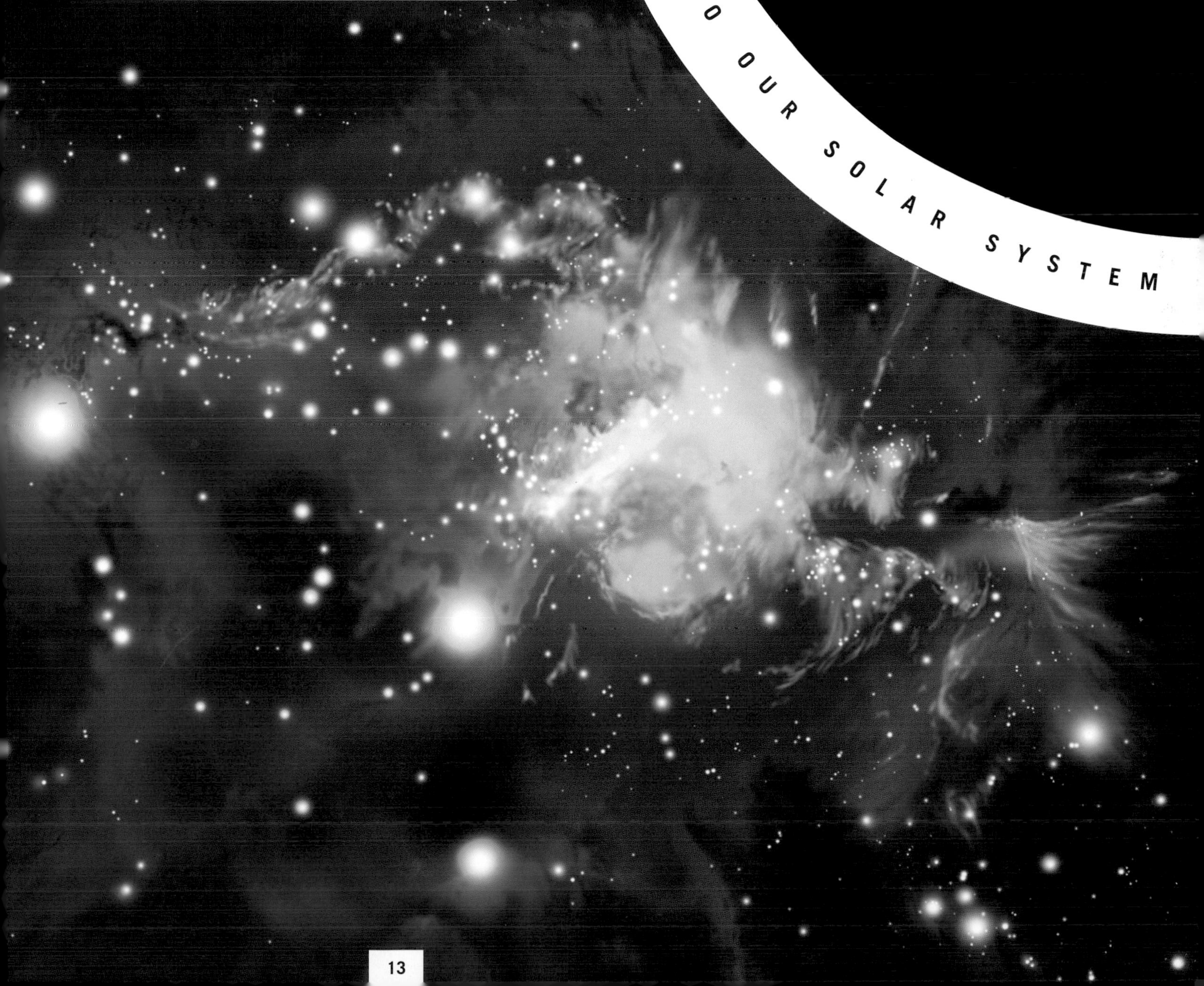

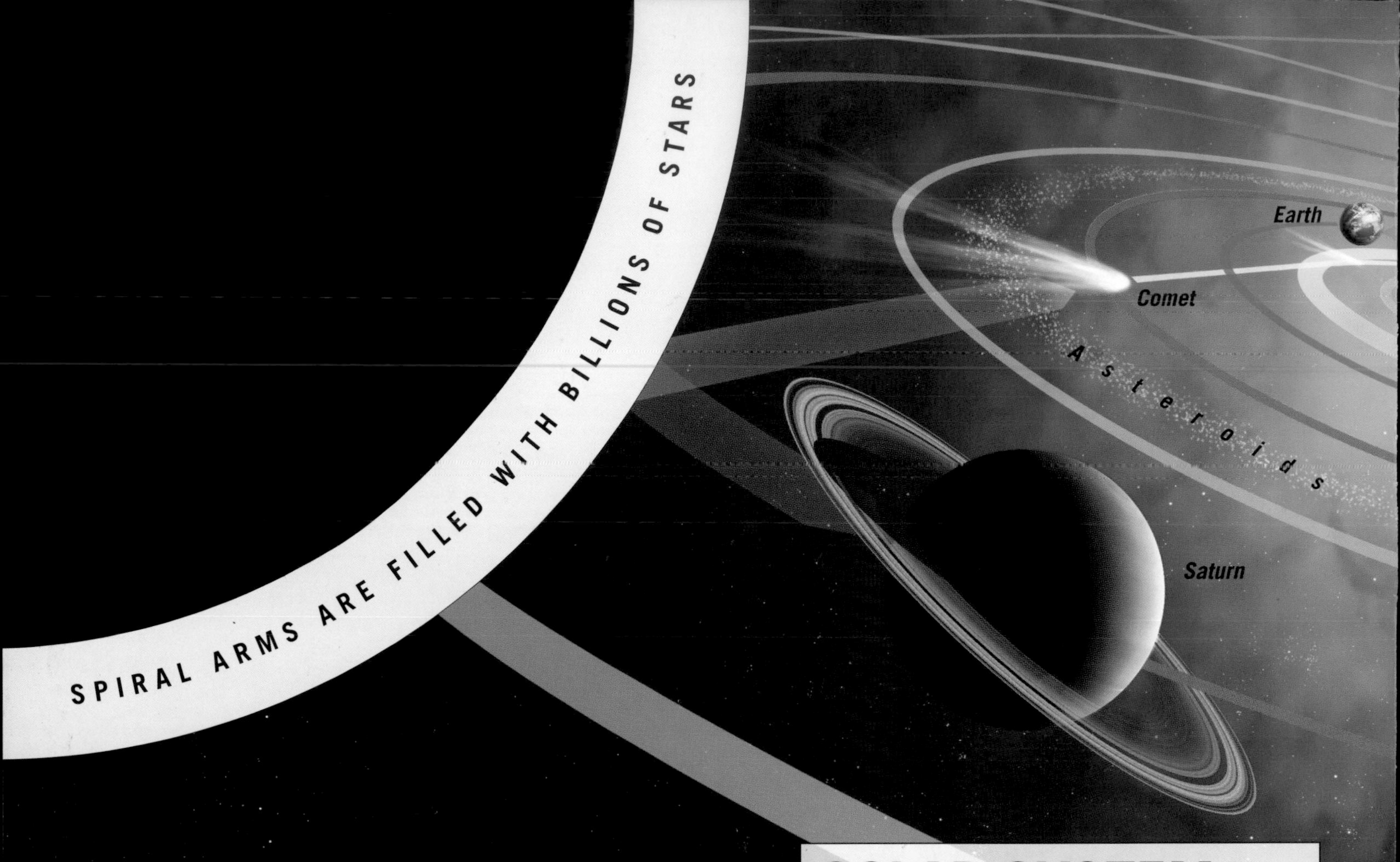

SPIRAL ARMS ARE FILLED WITH BILLIONS OF STARS

Earth

Comet

Asteroids

Saturn

# SOLAR SYSTEM

We have zoomed in on one of billions of stars in the Milky Way galaxy—the Sun *(see page 24)*. A relatively small star, it is orbited by its family of nine planets *(see page 26)*. The planets all travel around the Sun in a counterclockwise direction. The Sun and its planets make up the Solar System. Also included are the moons (which orbit the planets), asteroids, comets, meteoroids, and vast amounts of dust and gas. The Sun contains more than 99 percent of all the matter in the Solar System.

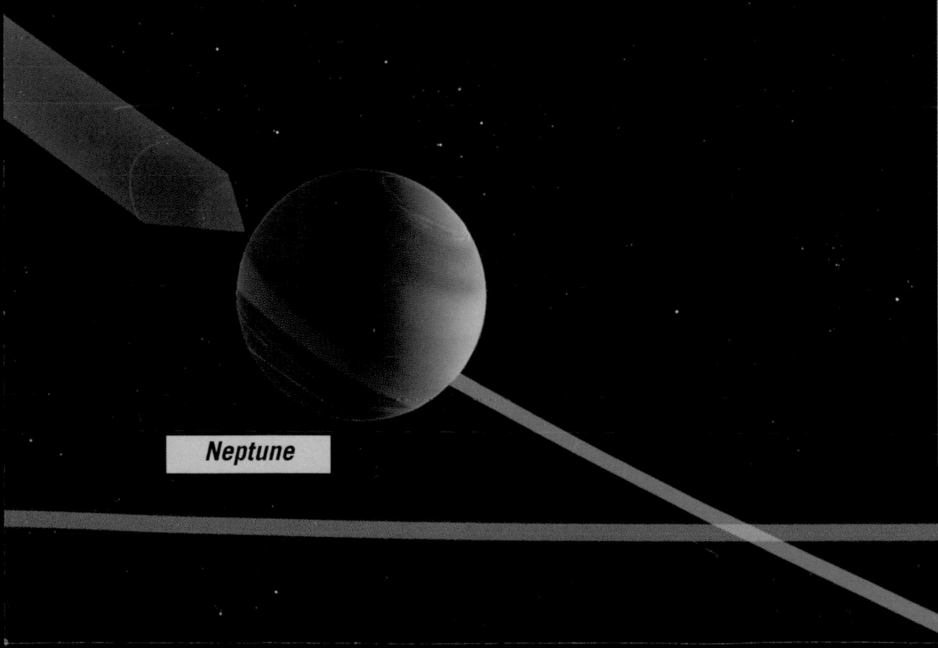

Neptune

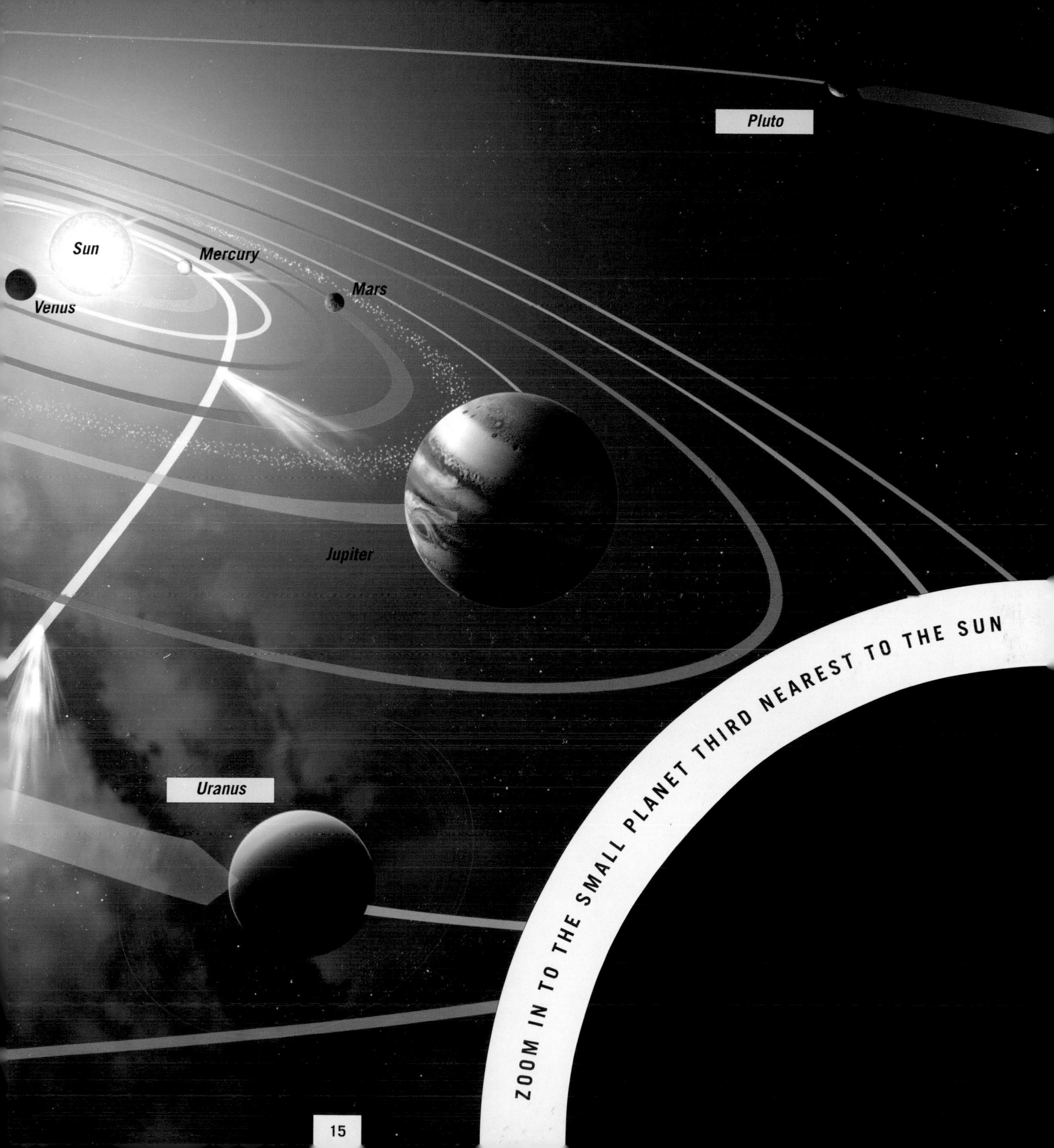
Pluto
Sun
Mercury
Venus
Mars
Jupiter
Uranus
ZOOM IN TO THE SMALL PLANET THIRD NEAREST TO THE SUN

# EARTH

One of the nine planets in the Solar System is, of course, our own planet Earth. It is the fifth-largest planet (although much smaller than the fourth largest, Neptune, *see page 27*). Earth is the only planet in the Solar System that we know to have life. Life on Earth is sustained by the presence of liquid water, which covers more than 71 percent of its surface. Earth's distance from the Sun provides a favorable temperature for its life-forms. Earth's atmosphere shields it from the Sun's harmful rays and protects it from bombardment by meteorites. Earth's oceans trap enough heat to avoid extremes of hot or cold.

EARTH IS A SMALL PLANET IN THE SOLAR SYSTEM

ZOOM DOWN TO SEE EARTH'S LAND SURFACE

LESS THAN ONE-THIRD OF EARTH'S SURFACE IS LAND

# LAND

Earth's landscape appears permanent and unchanging, but it is, in fact, changing all the time. Sometimes change happens quite quickly—for example, when part of a cliff falls into the sea, or when a volcano erupts. Usually, however, the change is a very slow process.

Earth's surface is divided into giant slabs, called plates. These plates slide around the

***Life is found nearly everywhere on Earth. Here, coniferous woodland grows on the upper mountain slopes. Roads, towns, pylons, and planted fields show the influence of people.***

***Observatory (houses telescope)***

globe so slowly that we cannot see them move. Plate movements may cause volcanic eruptions, earthquakes, and—where two plate edges collide—the crumpling up of land to form mountains.

Wind, rain, frost, rivers, glaciers, and the crashing of the waves all shape the landscape as well. They carve out wide valleys, gnaw away at cliffs, and eventually reduce mountain ranges to level plains. People can also change a landscape. Building reservoirs, diverting rivers, and quarrying a mountain are some ways humans change the land.

ZOOM DOWN TO THE GROUND TO SEE WHAT THE EARTH IS MADE OF

# ROCKS

Earth is one of five planets that are chiefly made of rock. Mercury, Venus, Earth, and Mars all have metal cores with rocky outer layers. Pluto has an icy surface and probably a rocky core. The gas giants—Jupiter, Saturn, Uranus, and Neptune—have relatively small rocky cores with thick gas or liquid outer layers.

There are three main types of rock in Earth's outer layer, called its crust. The first type is sedimentary rock. This is made from cemented rock fragments such as sand, gravel, mud, or the remains of living things. Igneous rocks are the second type. They are formed when molten rock (magma) from the interior rises, cools, and solidifies in Earth's crust. The third type is metamorphic rock. This is formed when rocks are subjected to great pressure and heat.

EARTH'S SURFACE IS DIVIDED INTO GIANT ROCK PLATES

*In many land areas, Earth's rocky crust has a covering of soil, a mixture of rocky fragments, and the rotting remains of plants and animals. Where there is soil, plants can grow.*

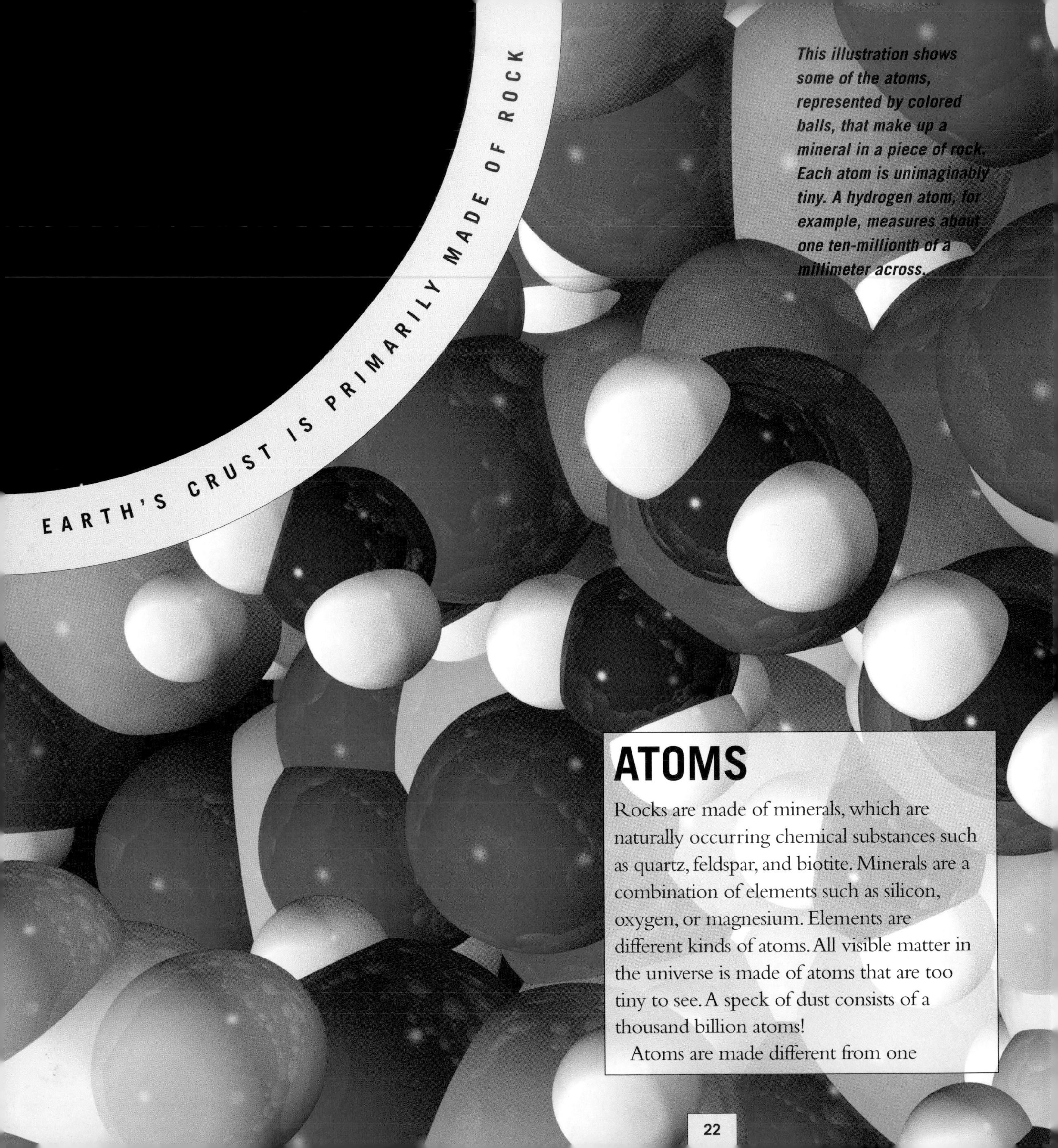

EARTH'S CRUST IS PRIMARILY MADE OF ROCK

*This illustration shows some of the atoms, represented by colored balls, that make up a mineral in a piece of rock. Each atom is unimaginably tiny. A hydrogen atom, for example, measures about one ten-millionth of a millimeter across.*

## ATOMS

Rocks are made of minerals, which are naturally occurring chemical substances such as quartz, feldspar, and biotite. Minerals are a combination of elements such as silicon, oxygen, or magnesium. Elements are different kinds of atoms. All visible matter in the universe is made of atoms that are too tiny to see. A speck of dust consists of a thousand billion atoms!

Atoms are made different from one

another by the number of particles they each contain. Different quantities of particles result in distinct elements. Hydrogen atoms are extremely light because they have just two particles. Lead atoms are heavy because they consist of many particles.

In a mineral, atoms of different elements are linked to one another, forming molecules. The molecules are packed very closely together, which is why minerals—and the rocks they make up—are dense, heavy solids.

# STARS

Stars are giant spinning balls of hot gases. Like massive nuclear power stations, they produce vast amounts of energy in the form of heat and light which they radiate across space as they shine. Stars vary enormously in size (Betelgeuse is 1,500 times the size of our star, the Sun) and the amount of light they give off. Some of the brightest stars emit more than 100,000 times the light of the Sun, while others are 100,000 times weaker.

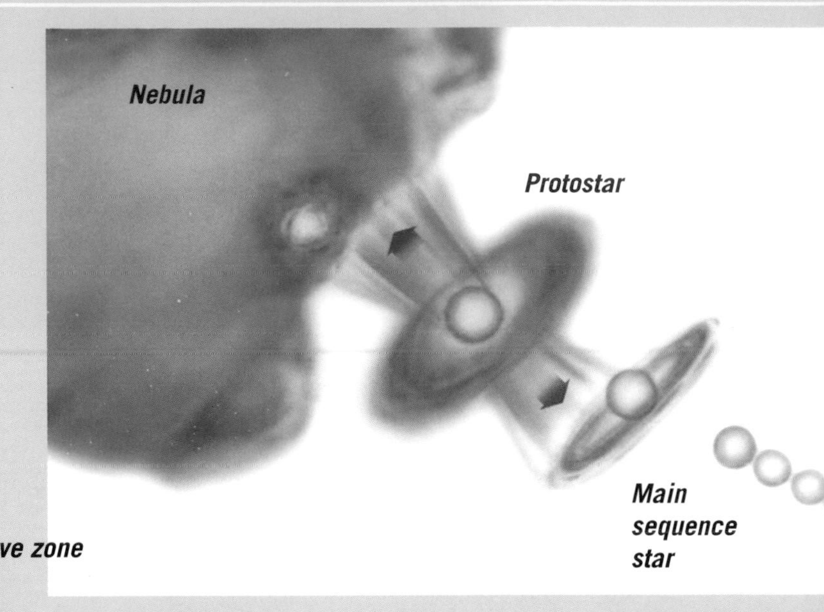


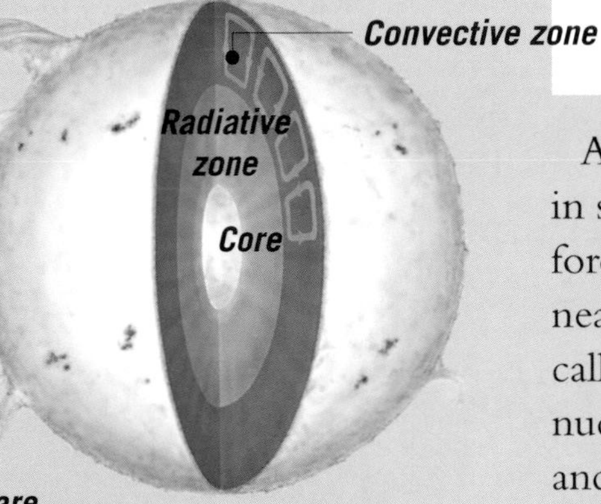

***At the Sun's center is the core, a region of immense pressure and heat 27 million°F (15 million°C). Energy produced at the Sun's core flows out through the radiative zone to the convective zone.***

A star begins its life when clouds of dust and gas in space, known as **nebulae**, compress under the force of gravity. Pressure from an old star exploding nearby may trigger this process. The resulting mass is called a **protostar**. Its core becomes so hot that nuclear reactions are triggered deep inside it. Gas and dust are blown away, although some may remain in the disc surrounding the star, eventually compressing to form planets.

The protostar is now a **main sequence** star. Most of these stars survive for billions of years on the power from nuclear reactions. Eventually, however, the fuel runs out, the core collapses, and the star swells into a **red giant**. A massive star will become a **supergiant** that will blast apart in a colossal explosion known as a **supernova**. It ends its days as a neutron star or a black hole *(see opposite)*.

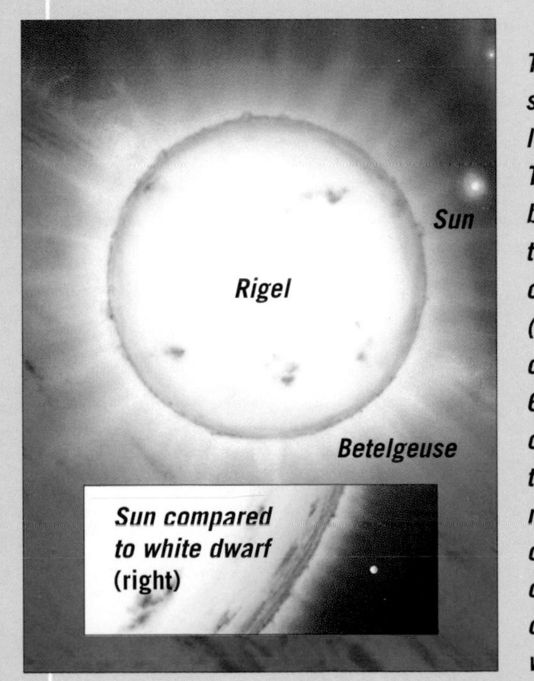


***The oldest stars are swollen red giants like Betelgeuse. They are much bigger and "cooler" than young blue ones like Rigel (both stars are part of Orion). Rigel is 60 times the size of our Sun, although the Sun is itself massive when compared to the collapsed core of an old star, known as a white dwarf.***

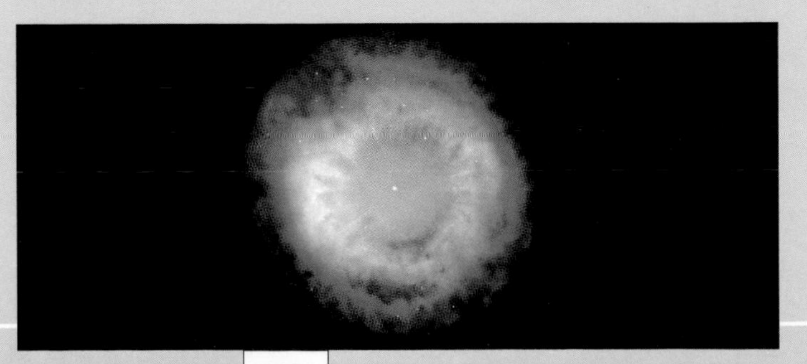

In a supernova, the exploding star shines brighter than 1 billion Suns! Chinese sky-watchers witnessed a supernova about 950 years ago. The Crab Nebula *(right)*, a cloud of gas in the shape of a crab, is all that remains of that—apart from a tiny, super-dense **neutron star** at its center.

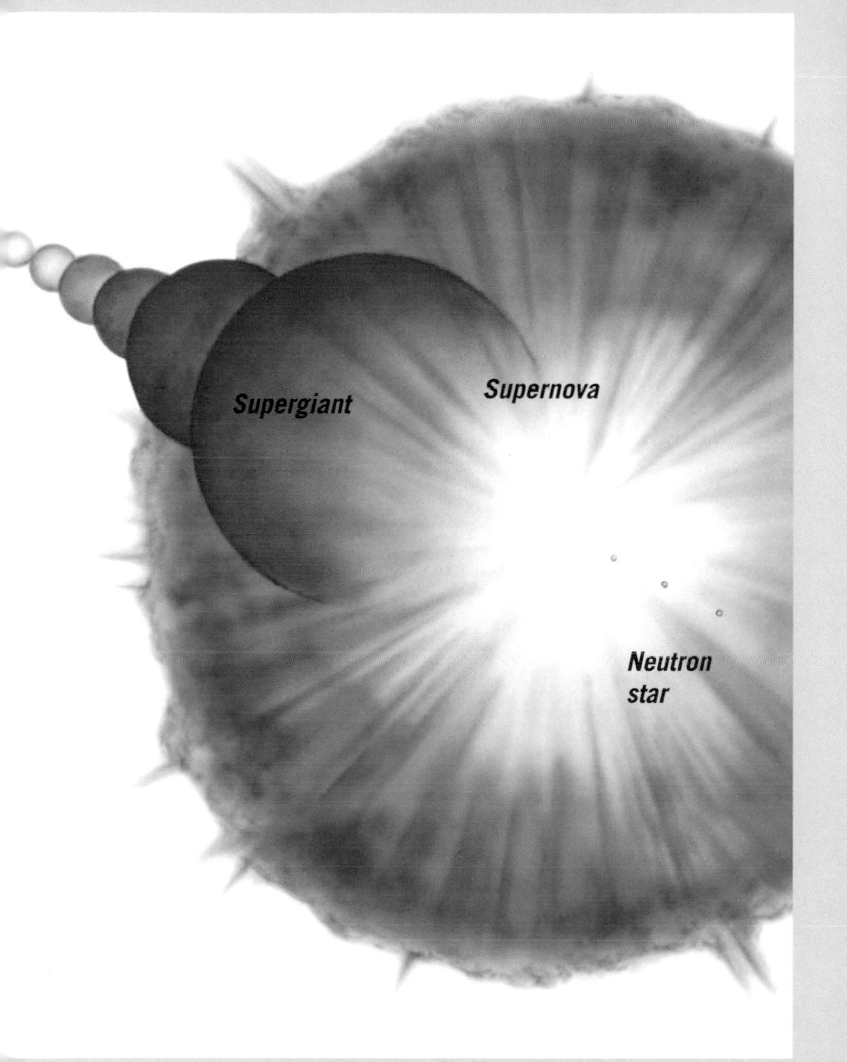


***When a less massive star (like our Sun, about 7 billion years from now) runs out of fuel, it also swells up—although not on the same scale as a supergiant. After a few million years, the outer layers flake off into space, leaving behind a ring of dust and gas, known as a planetary nebula. The Helix Nebula*** **(left)** ***is a planetary nebula. The collapsed core, no larger than a planet, is called a white dwarf.***

After a supernova, the old star's core may be so dense that it collapses in on itself. It shrinks to a tiny point surrounded by a region of space where gravity is so strong that nothing—not even light—can escape from it. Scientists call these points **black holes.** Anything lying close to a black hole disappears from the universe for good. Black holes are invisible, but it is possible to detect them. This blue star, for example, is being pulled around in a circle *(below)*. The gas torn from it forms a disc before plummeting into the black hole.

# THE PLANETS

Earth

Mars

Earth is the largest of the four inner planets: Mercury, Venus, Earth, and Mars. **Mercury**, nearest to the Sun, is heavily cratered. It has great extremes of temperature. **Venus**, permanently shrouded in thick clouds of deadly acid, has a surface temperature hotter than molten lead. Beneath the clouds, there are many volcanoes surrounded by lava plains.

**Mars** is the red planet, so-called because of the reddish iron oxide dust that coats its surface. Channels show signs of having been carved by running water, meaning that there may possibly once have been life on Mars.

The inner planets are dwarfed by the four gas giants—Jupiter, Saturn, Uranus, and Neptune. These planets consist chiefly of gas and have no solid surface. **Jupiter**, the largest planet, has patterns on its globe produced by high-speed winds and swirling storms. Its Great Red Spot is actually a giant storm—larger than Earth—that has been raging for at least 300 years.

All the gas giants have rings, but those of **Saturn** are by far the most spectacular. They are made of billions of blocks of ice and rock. **Uranus** is tilted 98 degrees from the vertical, meaning that it orbits the Sun almost on its side. **Neptune** also has high-speed winds racing around its globe.

Icy **Pluto**, the smallest, coldest, and outermost planet, has a very elongated orbit. It goes from just inside Neptune's orbit at one extreme, to billions of miles outside it at another.

Jupiter

| PLANET | DIAMETER | DAY measured in Earth days or hours | YEAR measured in Earth days or years | AVERAGE DISTANCE FROM SUN | SURFACE TEMPERATURE | MOONS |
|---|---|---|---|---|---|---|
| Mercury | 3,048 mi<br>4,878 km | 58.6<br>days | 88 days | 35.9 mil mi<br>58 mil km | -274 to +662°F<br>-170 to +350°C | none |
| Venus | 7,520 mi<br>12,103 km | 243<br>days | 225<br>days | 67.2 mil mi<br>108 mil km | 914°F<br>490°C | none |
| Earth | 7,926 mi<br>12,756 km | 23 hrs<br>56 min | 365.26<br>days | 93 mil mi<br>150 mil km | -94 to +131°F<br>-70 to +55°C | 1 |
| Mars | 4,222 mi<br>6,794 km | 24.6<br>hours | 687<br>days | 141.6 mil mi<br>228 mil km | -214.6 to +78.8°F<br>-137 to +26°C | 2 |
| Jupiter | 88,846 mi<br>142,884 km | 9.8<br>hours | 11.8<br>years | 483.6 mil mi<br>778 mil km | -238°F<br>-150°C | 28 |
| Saturn | 74,898 mi<br>120,536 km | 10.2<br>hours | 29.4<br>years | 886.7 mil mi<br>1,427 mil km | -292°F<br>-180°C | 30 |
| Uranus | 31,763 mi<br>51,118 km | 17.2<br>hours | 84 years | 1.8 bil mi<br>2,869 mil km | -346°F<br>-210°C | 21 |
| Neptune | 30,778 mi<br>49,538 km | 16.1<br>hours | 164.8<br>years | 2.8 bil mi<br>4,497 mil km | -364°F<br>-220°C | 8 |
| Pluto | 1,452 mi<br>2,324 km | 6.4 days | 248<br>years | 3.7 bil mi<br>5,906 mil km | -364°F<br>-220°C | 1 |

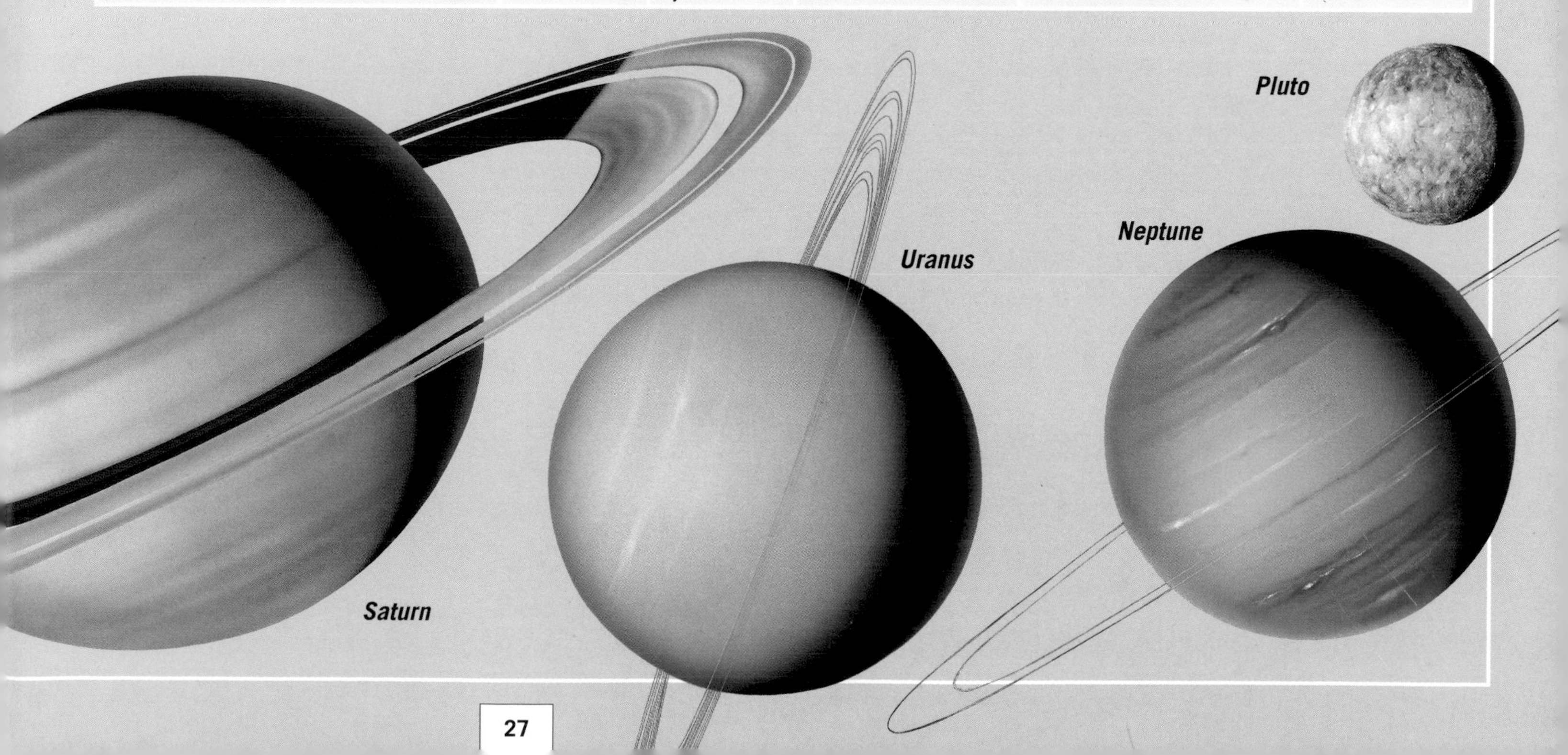

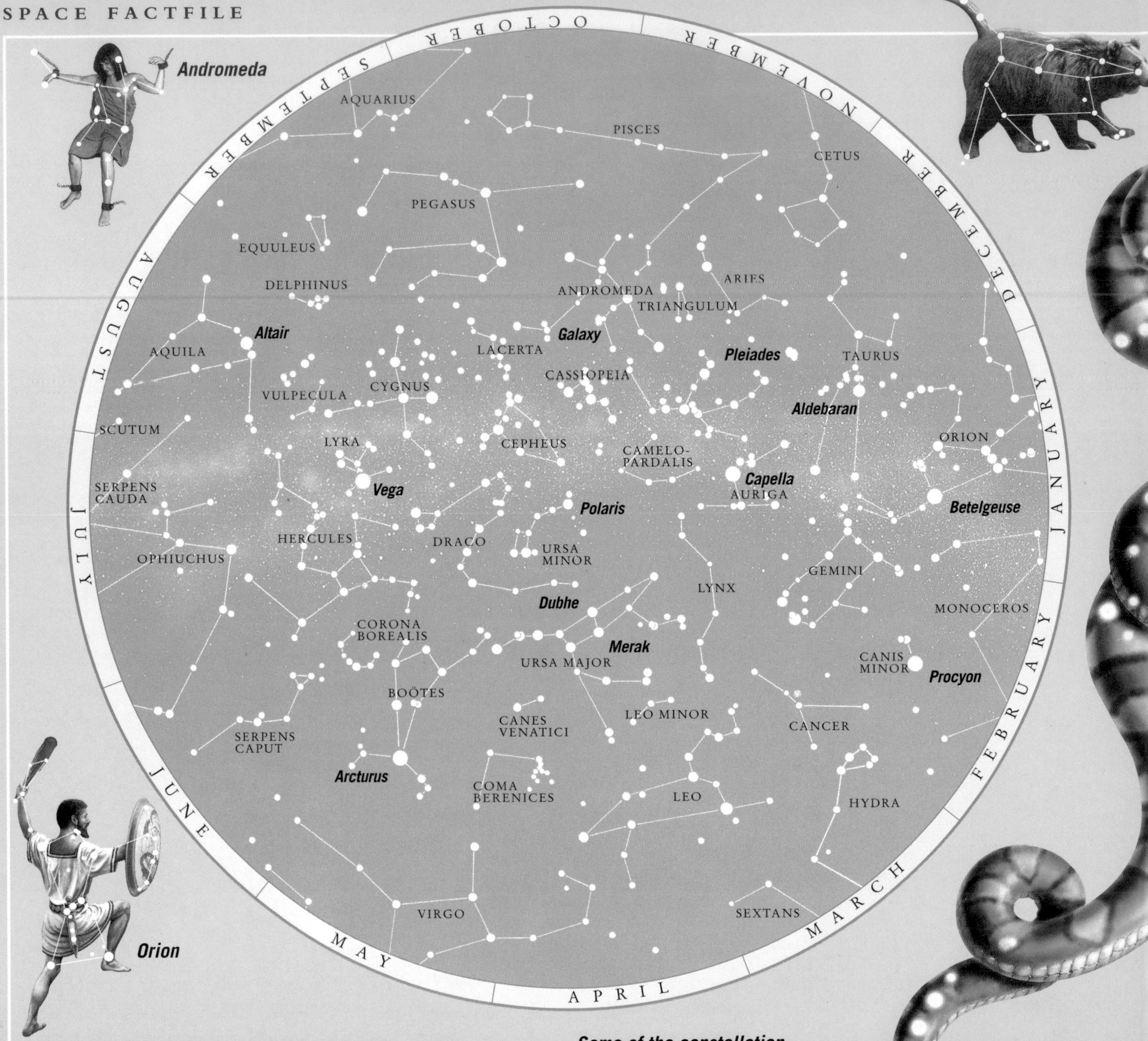


# CONSTELLATIONS

Centuries ago, astronomers grouped the stars they saw in the night sky into patterns. Observers imagined the starry arrangements looked like gods, heroes, and beasts from legends. We call these patterns constellations.

***Some of the constellation subjects are illustrated here. Orion is easy to spot: Three stars in a diagonal line form his belt (which points down to Sirius). Other stars make up Orion's dagger and shield.***

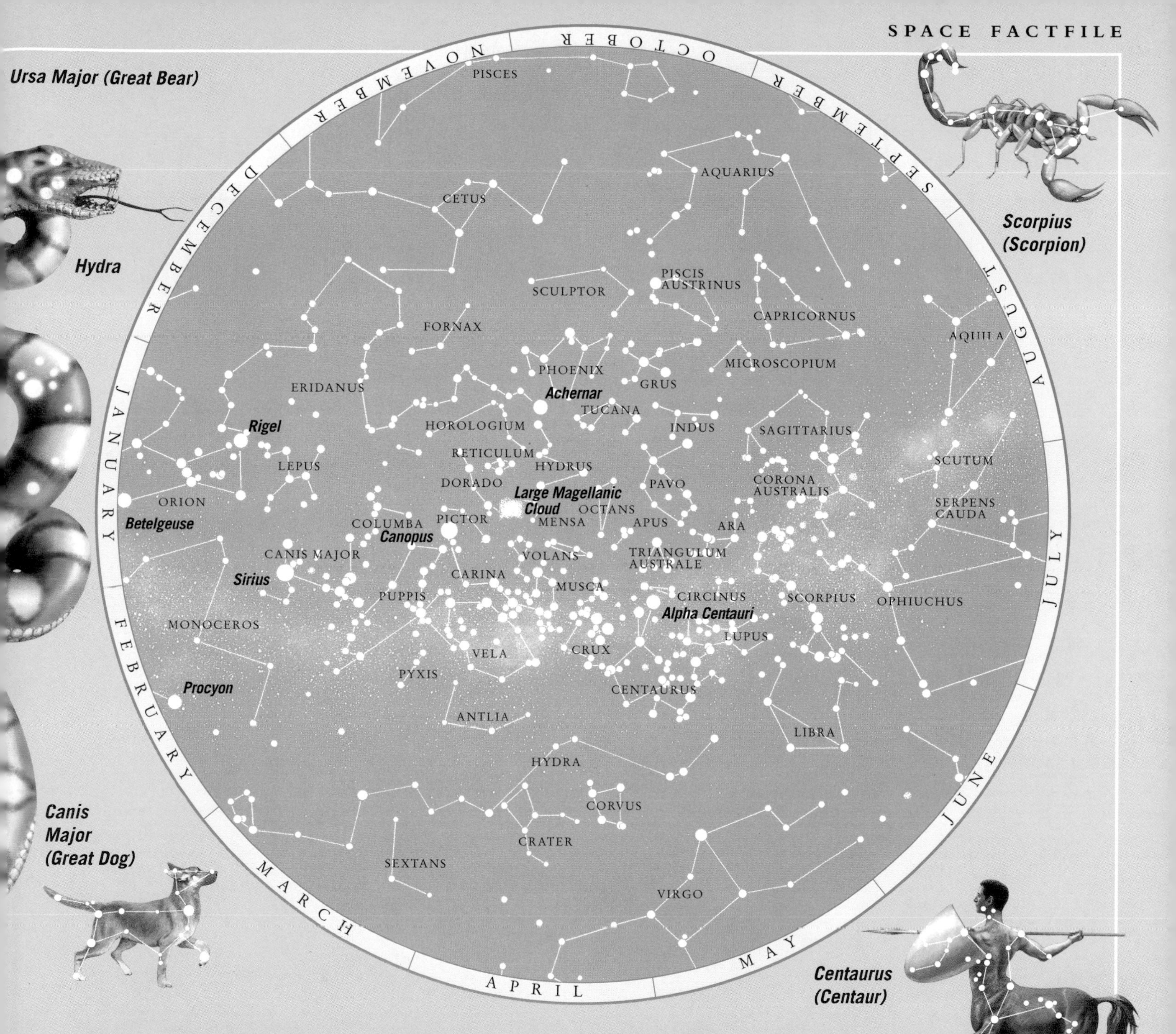


The 88 constellations include 48 known to the ancient Greeks. European explorers sailing the southern seas later added others visible only in southern hemisphere skies. Constellations are useful for locating certain stars and galaxies. Two of the brightest stars in Ursa Major, for example, point to Polaris, almost exactly due north.

To use these charts, turn the book around so the present month is at the bottom. Then, at 10:00 p.m., face south (northern hemisphere) or north (southern hemisphere) to locate the constellations on the chart.

# GLOSSARY

**Asteroid** A rocky body that orbits the Sun. Asteroids range in size from tiny specks to about 621 miles (1,000 km) in diameter.

**Atmosphere** The envelope of gases that surrounds a planet, moon, or star.

**Atom** A basic building block of matter. Elements—naturally occurring substances like hydrogen, carbon, or gold that cannot be broken down into simpler substances—are each made up from atoms of the same kind.

**Big bang** The origin of the universe, which took place in a gigantic explosion from an extremely hot and dense state about 15 billion years ago.

**Black hole** A region of space from which nothing, not even light, can escape. Its force of gravity is much more powerful than any normal star.

**Comet** An object made of dust and ice that orbits the Sun. On nearing the Sun, it develops two immensely long tails streaming away from the Sun.

**Constellation** A group of stars forming a pattern in the night sky.

**Galaxy** An enormous cluster of stars, planets, gas and dust. Galaxies may contain billions of stars. They are, themselves, gathered together in clusters of up to a few thousand.

**Light year** The distance that light, which moves at a speed of about 186,000 miles (300,000 km) per second, travels in one year. Astronomers use light years to measure the immense distances in space.

**Meteorite** A meteoroid that falls from space to land on the surface of a planet or a moon.

**Meteoroid** Rocks or dust particles that orbit round the Sun. Many meteoroids were once parts of asteroids. When a meteoroid burns up close to Earth it is known as a **meteor**.

**Molecule** A combination of atoms of different types bonded together. A molecule is the smallest part of a substance that can exist by itself and still possess its chemical properties.

**Moon** A smaller object that orbits a planet, also known as a natural satellite.

**Nebula** A cloud of gas or dust in space.

**Orbit** The circular or elliptical (oval-shaped) path followed by one object around another. For example, the Moon orbits Earth, while Earth orbits the Sun.

**Planet** A world that orbits a star. Planets do not radiate their own light, but reflect it from the star.

**Solar System** The Solar System consists of the Sun, together with the nine planets, their moons, comets, asteroids, meteoroids and a mass of gas and dust that all circle around it.

**Star** A globe of gas that produces heat from nuclear reactions inside its core and radiates light from its hot surface.

**Subatomic particle** The constituent parts of an atom. They include electrons and the protons and neutrons found in the atomic nucleus (center).

**Supernova** The massive explosion of a supergiant star.

**Tectonic plates** The large slabs into which Earth's surface is divided.

**Universe** All matter and space.